# Can we live on Mars?

Written by Giles Sparrow

Illustrated by El Primo Ramón

Foreword by Dr. Elizabeth Rampe, NASA

EARTHAWARE
KIDS

# A message from Dr. Elizabeth Rampe, Planetary Scientist, NASA

*Elizabeth is a planetary scientist. She studies rocks and minerals using orbiters and rovers on Mars to understand the planet's ancient past. She works with other scientists and engineers to help prepare for future robotic and human missions to the Red Planet.*

Scientists and engineers from around the world are preparing for one of the greatest challenges we have ever faced—the first human mission to Mars.

We have been learning about the Red Planet since it was first observed through a telescope more than 400 years ago. Since then, robotic missions to Mars have revealed a dry, cratered landscape and polar caps made of water and carbon dioxide ice. After years of study, we know that Mars was once a planet not unlike our own, with rivers, lakes, glaciers, and a thicker atmosphere. Ancient Mars had all the requirements for microscopic life, yet we have never found evidence for it there.

Landing humans on the Martian surface means we can continue this search for life. People can explore much faster than with a rover, and astronauts can study rocks they collect and bring back samples to Earth. The first human trips to Mars will be a gigantic leap in the exploration of our Solar System. We have already made short visits to the Moon, but the entire mission to Mars could take up to three years. Our explorations there could pave the way for the first permanent settlements off-world and indicate a new future for our species.

Human trips to Mars may happen sooner than we think. By using what you have learned from this book, practicing an interest in space at school, and using it in your future career, you too could one day help shape missions to Mars.

# Contents

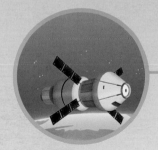
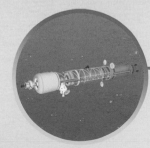

# Mind mapping

The reason this book is called Mind Mappers is because it is organized like a mind map. A mind map is a picture diagram that connects lots of different ideas. It is a very useful way to make complicated topics easy to understand. The mind map on this page looks at the question that is the title of this book—can we live on Mars? It divides the subject into the eight further questions, which are at the beginning of each chapter.

## Follow the lines

Find the question that you would like to explore and follow the colored lines to look at the individual topics. For example, there are two main ways we know about Mars— by looking at it through telescopes and by sending robot explorers. Keep following the lines to see how these topics subdivide.

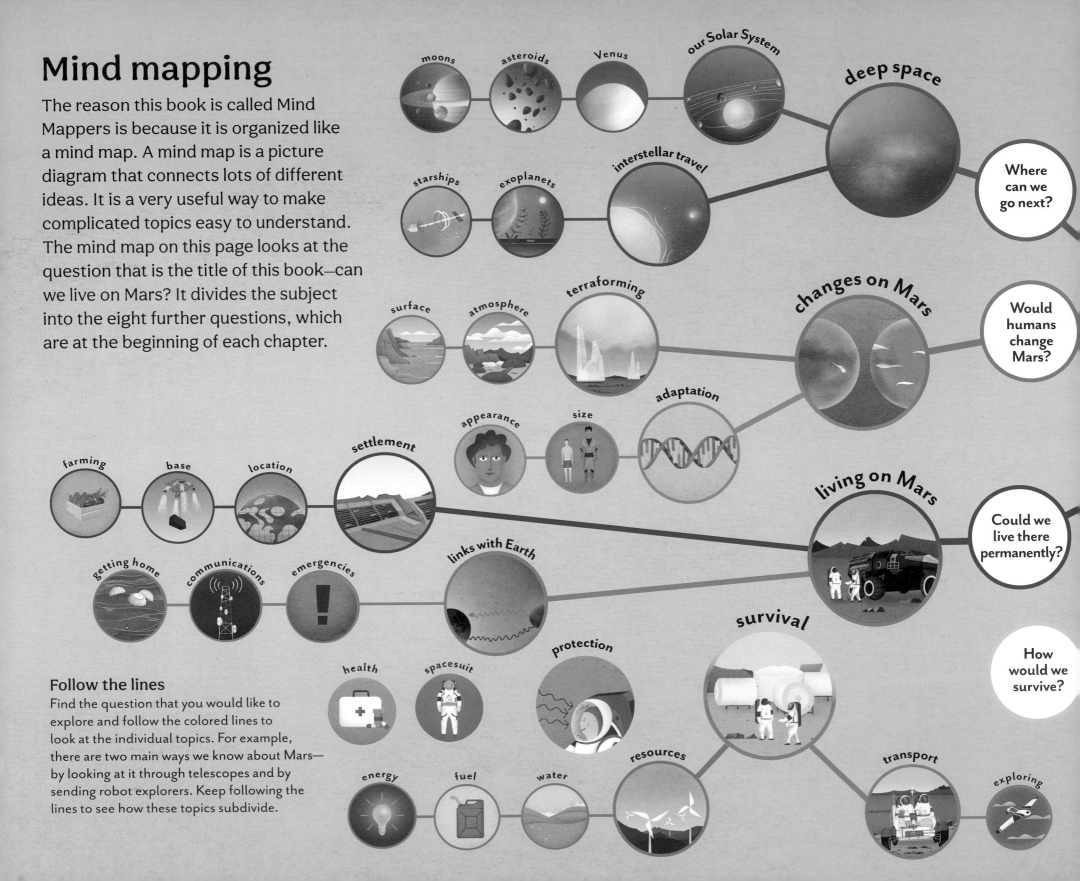

moons

asteroids

Venus

our Solar System

deep space

Where can we go next?

starships

exoplanets

interstellar travel

surface

atmosphere

terraforming

changes on Mars

Would humans change Mars?

appearance

size

adaptation

farming

base

location

settlement

living on Mars

Could we live there permanently?

getting home

communications

emergencies

links with Earth

survival

How would we survive?

health

spacesuit

protection

energy

fuel

water

resources

transport

exploring

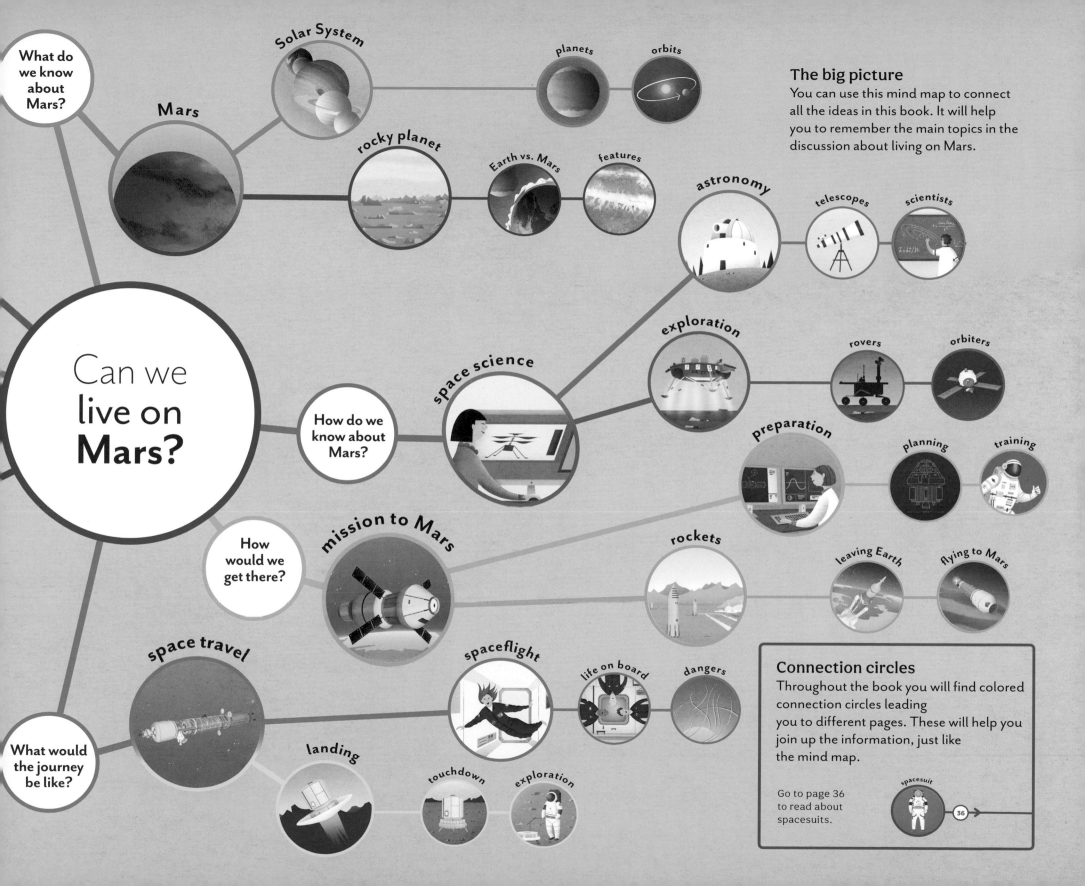

What do we know about Mars?

Mars

Solar System

planets

orbits

rocky planet

Earth vs. Mars

features

astronomy

telescopes

scientists

The big picture

You can use this mind map to connect all the ideas in this book. It will help you to remember the main topics in the discussion about living on Mars.

Can we live on Mars?

How do we know about Mars?

space science

exploration

rovers

orbiters

preparation

planning

training

How would we get there?

mission to Mars

rockets

leaving Earth

flying to Mars

space travel

spaceflight

life on board

dangers

What would the journey be like?

landing

touchdown

exploration

Connection circles

Throughout the book you will find colored connection circles leading you to different pages. These will help you join up the information, just like the mind map.

Go to page 36 to read about spacesuits.

spacesuit

36

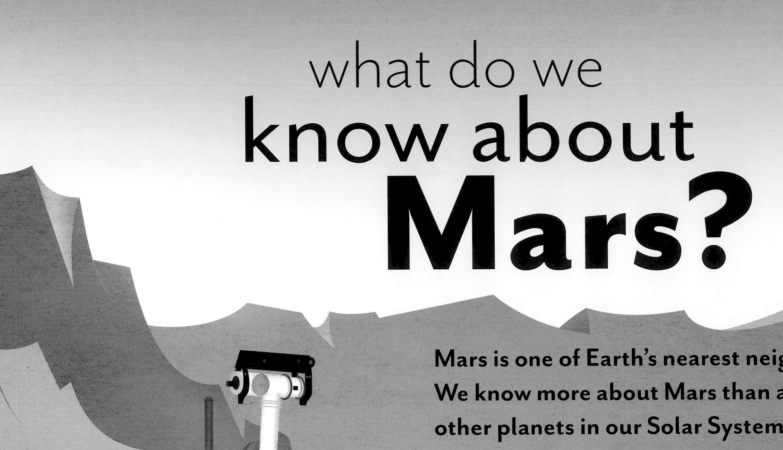

# what do we know about Mars?

Mars is one of Earth's nearest neighbors. We know more about Mars than any of the other planets in our Solar System, aside from our own. The similarities between Mars and Earth as well as their closeness make Mars an ideal target for future human explorers.

# Mars

Mars is a small red planet about half the size of Earth. It is not the nearest planet to ours, but it has more in common with Earth than any other nearby planet.

## Solar System

Mars is the fourth planet from the Sun in the Solar System, the collection of planets that travel round the Sun.

## rocky planet

Mars has a rocky surface, with a thin atmosphere and weak gravity.

planets

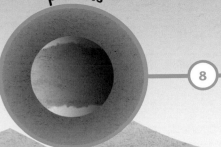

8

orbits

9

Earth vs. Mars

10

features

12

# Where is Mars?

Mars is the small red planet shining in the night sky. We can see it because it is one of the closest planets to Earth. Mars is the fourth planet from the Sun, and the only one we have landed robot explorers on. While it is dry, dusty, and possibly lifeless, it is still the most Earthlike of all the other planets in our Solar System, so is the best place for human exploration in the near future.

## Inner Solar System

Between the asteroid belt and the Sun lie four rocky planets. They have very different surface environments depending on their size and how far they are from the Sun.

## The Solar System

There are eight planets in the Solar System. The planets all orbit the Sun, along with many smaller rocky and icy objects.

### Mercury

The smallest planet is closest to the Sun. One side of its surface is scorching and airless, and the other is icy cold.

### Venus

Almost twice as big as Mars, Venus has a thick, choking atmosphere that is deadly to life as we know it.

### Earth

The biggest of the inner planets is Earth. It has a protective atmosphere, watery oceans, and bursts with life.

### Mars

The Red Planet is the inner planet farthest from the Sun. This, its small size, and thin atmosphere make Mars drier and colder than Earth.

# Outer Solar System

Beyond the asteroid belt are four big planets made of gas or ice. Each of these giant planets has a large number of smaller worlds, called moons, orbiting around them.

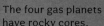
The four gas planets have rocky cores.

## Neptune
The outermost planet has a stormy atmosphere and some of the fastest winds in the Solar System.

## Uranus
The blue-green giant planet Uranus has an icy interior and a system of thin rings.

## Saturn
A bright system of rings surround the second-biggest planet in the Solar System. Their countless tiny pieces of ice sparkle as they reflect sunlight.

## Jupiter
The biggest planet in the Solar System is on the far side of the asteroid belt. Heat from inside Jupiter powers the weather in its colorful, stormy atmosphere.

discovering Mars

16 →

## Asteroid belt
A region of small, scattered rocks called asteroids divides the inner and outer Solar System.

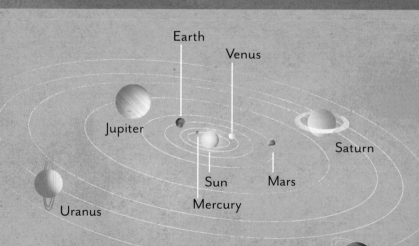

Earth

Venus

Jupiter

Sun

Mercury

Mars

Saturn

Uranus

Neptune

## Orbits
Planets travel around the Sun in roughly circular paths called orbits. The inner planets are much closer together than the outer planets. Beyond Mars, the gaps between the planets get bigger and bigger.

### Mars connections
Mars is a small, rocky planet. It is the fourth planet from the Sun in our Solar System. Mars is the most Earthlike planet near us, making human missions there a possibility.

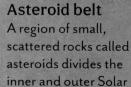

# Comparing Earth and Mars

Mars is the planet in our Solar System most like Earth, but it is still very different. Earth is bigger and is closer to the Sun, and has plenty of surface water and a thicker atmosphere. These have made it an ideal home for a huge variety of life. Mars's smaller size and greater distance from the Sun make it cold, dry, and probably lifeless. Yet it also has a stunning landscape of towering volcanoes and deep canyons much bigger than any on Earth.

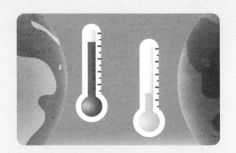

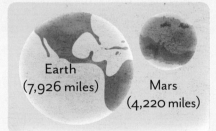

Earth
(7,926 miles)

Mars
(4,220 miles)

### Temperature
Earth has freezing poles and hot deserts, but its average temperature is a mild 60°F. On Mars, it rarely gets above freezing, averaging -85°C.

### Size
Earth is about twice as wide as Mars, and weighs about ten times more. It would take more than six Marses to fill the Earth.

atmosphere

### Life on Earth
Earth is home to many different types of life. Plants make the land look green from space. They help to make air that animals can breathe.

### Gravity on Earth
A force called gravity pulls everything toward Earth's surface. It keeps your feet on the ground and helps stop Earth's atmosphere escaping into space.

# Big differences

We do not know of any liquid water or life on Mars. These conditions mean changes on the surface can take thousands or even millions of years. Earth's water, atmosphere, and life work together to create a planet that is always changing.

## Atmosphere

Martian air is much thinner than Earth's atmosphere. It is mostly made of carbon dioxide gas, which humans cannot breathe.

**Olympus Mons**
Mars is home to the biggest mountain in the Solar System. Olympus Mons is twice the height of Mount Everest.

**Tharsis Montes**
Three giant extinct volcanoes sit on a huge bulge in the Martian surface called Tharsis.

**Valles Marineris**
Cracks three times deeper than the Grand Canyon wrap around one side of the planet. Valles Marineris is long enough to stretch across North America.

## Life on Mars

A lack of liquid water and dangerous rays from the Sun make it difficult for life to survive on Mars's surface. But scientists think things might have been different in the past, which is why we are searching for signs of life on Mars.

Phobos

Deimos

## Martian moons

Mars has two moons, called Phobos and Deimos. They are much smaller than Earth's Moon, but because they orbit close to Mars they appear large in the Martian sky.

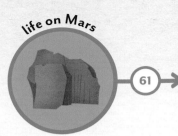
life on Mars

61

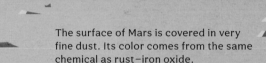
Curiosity

The surface of Mars is covered in very fine dust. Its color comes from the same chemical as rust—iron oxide.

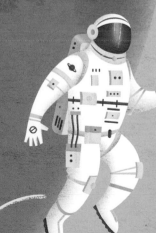

## Gravity on Mars

Mars's weaker gravity means you weigh three times less than on Earth, so you could jump much higher.

**Mars connections**

Mars's smaller size, thin atmosphere, and position farther from the Sun give it very different conditions from Earth's. They make it difficult for humans to survive on Mars, so we would need to plan a mission there very carefully.

# Conditions on Mars

The temperature, atmosphere and gravity of Mars are very different from Earth's. But in other ways the two planets are very alike. Both have rocky surfaces, meaning we can land rovers and people there. They have similar day lengths and seasonal patterns. Mars also has plenty of water ice that future astronauts could use for water and rocket fuel.

The south tilts toward the Sun, so it is summer there, and winter in the north.

## Surface

The rocky Martian landscape has many similarities to Earth—both have valleys, mountains, and volcanoes. Underground tunnels, carved by lava from ancient volcanoes, might be an ideal location for a base on Mars.

## Cycle of seasons

Mars spins on a slight tilt. Like Earth, this causes it to have four seasons. Mars takes 687 days to go around the Sun, so its seasons last about twice as long as Earth's.

lava tunnels

46

## Rocky planet

Mars and Earth are both made from rock and metal. Mars has lost more of its internal heat because of its small size compared to Earth, so does not have the energy to power volcanoes or strong earthquakes.

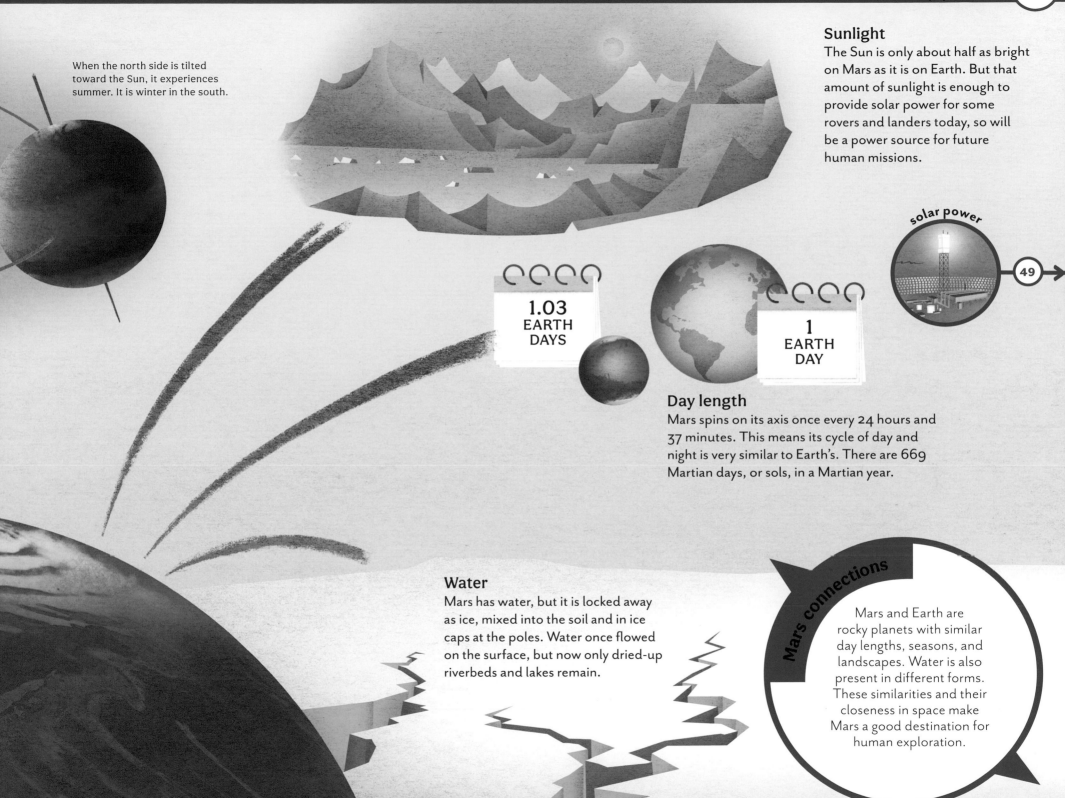

When the north side is tilted toward the Sun, it experiences summer. It is winter in the south.

## Sunlight

The Sun is only about half as bright on Mars as it is on Earth. But that amount of sunlight is enough to provide solar power for some rovers and landers today, so will be a power source for future human missions.

solar power

49

1.03 EARTH DAYS

1 EARTH DAY

## Day length

Mars spins on its axis once every 24 hours and 37 minutes. This means its cycle of day and night is very similar to Earth's. There are 669 Martian days, or sols, in a Martian year.

## Water

Mars has water, but it is locked away as ice, mixed into the soil and in ice caps at the poles. Water once flowed on the surface, but now only dried-up riverbeds and lakes remain.

## Mars connections

Mars and Earth are rocky planets with similar day lengths, seasons, and landscapes. Water is also present in different forms. These similarities and their closeness in space make Mars a good destination for human exploration.

# how do we
# know about
# Mars?

Even at its closest to Earth, Mars is 34 million miles away and a tiny red dot in our night sky. Scientists learn about Mars by using telescopes so they can see the planet in more detail. They also send robot explorers to visit the planet on our behalf.

# space science

Our knowledge of Mars comes from the work of space scientists. Astronomers measure Mars from Earth, while engineers build robot probes to visit it.

## astronomy

Astronomers use telescopes to view Mars close up and understand more about it. Scientists can use this information to learn about its history.

## exploration

Space probes carry scientific instruments to Mars. They photograph its surface from orbit, or land to take measurements and close-up images.

### telescopes

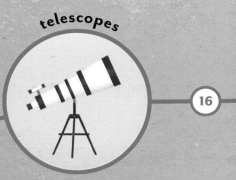

16

### scientists

17

### rovers

18

### orbiters

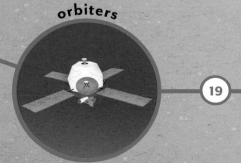

19

# Discoveries from Earth

Mars is one of a handful of bright planets that move across the night sky, while the stars stay fixed in place. For many centuries, stargazers could only see Mars with the naked eye, so could learn little about it. After the telescope was invented around 1608, astronomers could finally see Mars in more detail—as a tiny red globe with marks on its surface. Until the 1960s, telescopes were the only way of studying Mars.

**2000 BCE**
Egyptian stargazers realize Mars is a kind of "wandering star" called a planet. They believe it is a messenger of the falcon-headed god Horus.

## Learning about Mars

Scientists and stargazers have made many discoveries about Mars from Earth. The biggest ones have come by using telescopes. They have shown that Mars has some things in common with Earth, but that it is also a very different place.

robot discoveries

18

moons

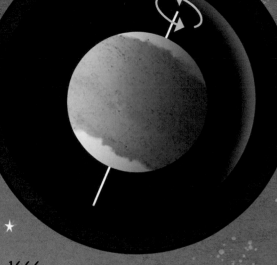

**1666**
Giovanni Cassini uses a telescope to spot white ice caps at the poles of Mars. They grow in winter and shrink in summer.

**1784**
William Herschel sees changing clouds above the surface of Mars. He wonders if there might be Martians living on the planet.

**1877**
Using a powerful telescope, Asaph Hall discovers two small moons orbiting Mars. These are named Phobos and Deimos.

$CO_2$

$CO_2$  $CO_2$  $CO_2$  $CO_2$

$CO_2$

$CO_2$  $CO_2$

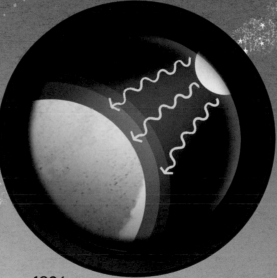

**1926**
Walter Sydney Adams works out that Mars has a very thin atmosphere, with no oxygen or water. This means the planet must be a cold, dry desert.

**1947**
Gerard Kuiper shows that the air on Mars is mostly made of carbon dioxide gas, meaning Mars is unlikely to support life.

**Mars connections**

Telescopes allow people to study Mars from a great distance and find out more about it. These discoveries have shown how hard it will be for astronauts to live there.

# Robot explorers

For sixty years, humans have been sending robot space probes to explore Mars up close and send information back to Earth. Orbiters circle Mars to photograph and measure it from space, while landers touch down to tell us more about the Martian surface. Some landers carry rovers–wheeled robots that roll across the land, performing experiments to learn more about Mars's past and present.

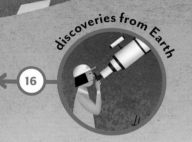

← 16 discoveries from Earth

## Exploring the surface

Rovers use lots of computers, cameras, and special instruments to study the Martian rocks, soil, and climate. Their wheels each have separate motors to help them avoid getting stuck as they drive around.

## Curiosity and Perseverance

A pair of near-identical rovers are the biggest vehicles to explore Mars so far. Curiosity drills into and analyzes rocks and Perseverance collects rock samples. Both have a long robot arm to take close-up photos.

## Rovers and landers

Landers are robot science labs that stay in one place and measure conditions around them. Rovers can explore more widely, collecting samples and photographing their surroundings.

### First landers

The two Viking landers touched down in 1976. They carried tools to measure Martian weather and search for life.

### First rover

The 1997 Pathfinder mission took the first rover to Mars—a small, wheeled robot called Sojourner.

# Exploring from space

Many robot explorers visiting Mars stay in space. Some fly past the planet, while others go into orbit. These can study the whole planet rather than just a small area. They make maps of the surface, measure the atmosphere, and can even detect chemicals in the rocks.

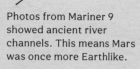

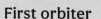

Photos from Mariner 9 showed ancient river channels. This means Mars was once more Earthlike.

## 3D views
The European orbiter Mars Express photographs the surface from different angles. It puts these together to create stunning 3D views of the planet.

### Mars connections
Robot spacecraft are sent to Mars to learn more about it. Orbiters photograph the planet and can spot good sites for human missions. Landers study the conditions and can test technology that people will need to explore Mars.

## First orbiter
The first spacecraft ever to orbit another planet was Mariner 9. It arrived at Mars in 1971 and sent back detailed pictures of the planet from space.

## Flying visit
Mariner 4 flew past Mars in 1965 without stopping. It sent back the first images ever taken of another planet. The close-up pictures of the surface showed a cratered, lifeless world.

### Landing on Mars

32

## Looking for water
The twin rovers Spirit and Opportunity were designed to study water on Mars. They found signs that it once flowed on the surface.

## Exploring the north
In 2008, the Phoenix lander touched down in a nothern region of Mars. It found water ice just below the surface using a scoop on its arm.

## Studying the soil
China's first Mars rover landed in 2021. It spent a year examining the surface soil and Martian atmosphere.

# how
# would we
# get there?

We will need a very powerful rocket to get to Mars. The farthest humans have ever traveled so far is to the Moon, and Mars is at least 130 times farther away than that. A trip to Mars will require lots of fuel, and many years of preparation and training.

# mission to Mars

A mission to Mars will be risky and take many months. Astronauts will need a spacecraft that protects them and provides comfortable living conditions.

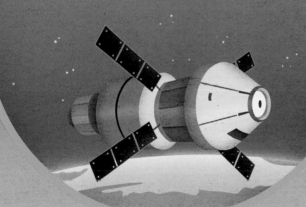

## preparation

Getting to Mars involves years of work to plan the trip and develop the spacecraft. The crew will also need months of training.

## rockets

The vehicles that take the crew from Earth to Mars will be powered by rockets— giant, powerful engines.

### planning

22

### training

24

### leaving Earth

26

### flying to Mars

27

# When to go

Picking the best time to set off for Mars makes a big difference to the length of the journey. Every so often, Earth and Mars get closer together. As part of their planning for a Mars mission, scientists and engineers have to work out when this will be. They also need to decide how long astronauts will stay on Mars for. Should they head back to Earth after a brief visit, while the planets are still quite close together, or should they settle in for a long wait until the planets are next close together?

TRIP IS UP TO
9 MONTHS LONG

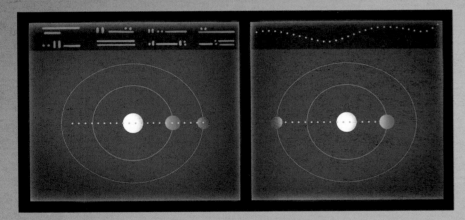

### Close together
Earth and Mars orbit the Sun at different speeds, but they draw close to each other roughly every 26 months. This event is called opposition, and is the best time for a trip to Mars.

### Launch
Mars moves along its orbit at tens of thousands of miles an hour, so a spacecraft from Earth cannot just be pointed in a straight line at Mars. Instead, it must launch toward the spot where Mars will be in several months' time.

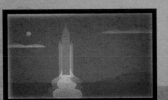

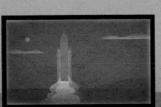

rocket launch

26

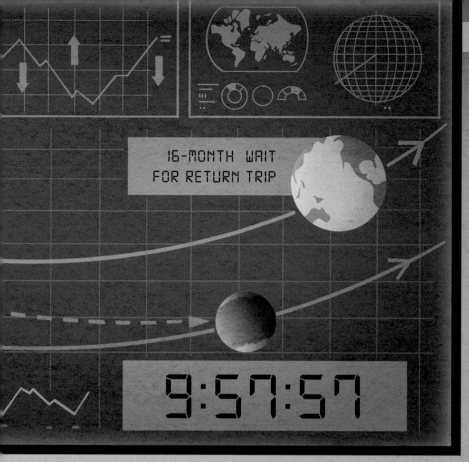

16-MONTH WAIT FOR RETURN TRIP

9:57:57

Depart for Mars from an orbit around the Moon.

NEXT OPPORTUNITIES TO GO

2035    2050    2067

## Arrival
A Mars-bound spacecraft has to time its arrival at the planet perfectly, or it may miss Mars completely! The spacecraft will then need to change its speed and slow down so it can enter orbit around the Red Planet.

## Best times to go
Mars's orbit around the Sun is more egg-shaped than a circle, so Mars can get a long way from Earth. This means the journey time from Earth to Mars can vary a lot. The best launch windows happen when Mars is closer to the Sun as well as the Earth.

arriving at Mars

## Mission Control
Astronauts on a trip to Mars will rely on hundreds of people planning and monitoring every part of the mission. Most key events in the journey will be planned long before takeoff.

## Mars connections
It is important to know when Earth and Mars will be close to each other. This helps keep the journey shorter, which means missions cost less money and need less fuel. It also reduces health risks to astronauts making the trip.

# Preparation and training

Astronauts go through months of training to prepare for a trip in space, but the mission to Mars will take years of preparation. Engineers, scientists, doctors, and volunteers will test and practice every part of the journey to make sure everything is as safe as possible. Astronauts will train with all the different systems and environments they may face on the voyage through space and on the surface of Mars.

## Space habitats

During a Mars mission, people will have to live together in cramped spaces, where they cannot go outside whenever they want. Volunteers spend months in similar conditions on Earth to help scientists better understand how to cope.

Mars surface

12

## Learning to explore

Although nowhere on Earth is exactly like Mars, dry deserts are the most similar. Here, astronauts can rehearse in fairly Mars-like conditions. Spacesuit designs can be tested for the tasks that astronauts will need to do on Mars.

## Virtual reality training

When they cannot practice on Earth, astronauts can use virtual reality. A headset shows the wearer images that help them imagine they are on Mars's surface, while sensors pick up their movements.

## Flight simulators

Pilots for a Mars mission will use flight simulators to rehearse launch, landing, and other key events. They do this hundreds of times so they are ready for every situation.

## Weightlessness

There is no gravity in space, so astronauts will be weightless on the journey to Mars. On Earth, they can get used to this inside aircraft that make steep dives toward the ground. This causes everything on board to float as if weightless.

## Spacewalks

Astronauts can train for work in low or zero gravity in deep swimming pools. The water supports their spacesuits so they float as if they were in a weightless environment.

### Mars connections

The crew will be far from Earth's help for most of their trip. They need to be well trained in every part of their mission. Scientists, engineers, and volunteers must test all the equipment so they know it is safe.

spacesuits

24

training

# On a rocket

The best way to get to Mars is by using a rocket. Rockets send spacecraft carrying astronauts into space. Most rockets work by burning a store of chemicals called propellants. This produces hot gases, which rush out of the back of the rocket and blast the spacecraft in the opposite direction. Because rockets do not need to burn air, they can work in many different conditions, including on Earth and Mars, and in the emptiness of outer space.

The launch-abort system rocket can pull the spacecraft to safety in emergencies.

launch vehicle

The spacecraft sits on top of the launch vehicle.

booster rocket

spacecraft

upper-stage rocket

stage adapter

core-stage rocket

### Getting off Earth
Rockets are the only way of getting into space. By quickly burning lots of fuel, a rocket-powered launch vehicle is pushed off the ground, gaining speed until it reaches space.

### Into orbit
A launch vehicle is a system of rocket engines used to carry a spacecraft into orbit. The rockets make sure the spacecraft is moving so fast that gravity cannot pull it back to the ground.

booster rocket

### Different stages
Launch vehicles are built from several "stages". Each stage has its own rocket engine. The stages fire together, or one after another. When a stage runs out of fuel it falls away, reducing the vehicle's weight.

The orange core-stage and booster rockets fire together to push the launch vehicle off the ground.

Once the booster rockets run out of fuel, they separate and fall away.

space travel **30**

## Lunar Gateway station

The crew on their small spacecraft arrive at a Lunar Gateway space station, which will be built in space and orbit around Earth's Moon. The crew transfer to the space station ready for their onward trip to Mars.

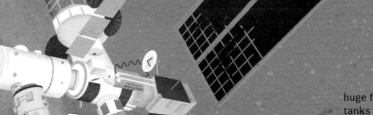

huge fuel tanks

## Building the spacecraft

A large spacecraft takes the crew the rest of the way to Mars. This Mars Transfer Vehicle will be assembled using parts brought from Earth.

Mars Transfer Vehicle

the crew's living quarters

## Astronauts on board

The crew transfer from the Lunar Gateway space station to the Mars Transfer Vehicle. The huge spacecraft's engines will launch it from lunar orbit on a course for Mars.

### Mars connections

Powerful rockets are needed to get off Earth and into orbit. Rockets are also used when a spacecraft has to change speed or direction. A different vehicle takes astronauts from orbit around Earth all the way to Mars.

The launch-abort system separates from the top of the spacecraft.

The spacecraft and upper-stage rocket separate from the used-up core stage.

The spacecraft opens its solar panels to make power.

The upper-stage rocket falls away as the spacecraft leaves Earth's orbit and continues into space.

# what would the journey be like?

The journey to Mars will take several months. During this time, astronauts will face several challenges, from eating and sleeping in weightless conditions to looking after their health. They will then have to face a rapid descent through the Martian atmosphere.

# space travel

Humans have never traveled so far in space and for so long. Their spacecraft will carry everything they need for the journey, including food, equipment, and rocket fuel.

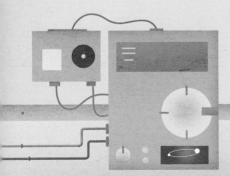

## spaceflight

Astronauts will have to adapt to life in long-term zero gravity. They must also be wary of the risks of spaceflight.

### life on board

30

### dangers

31

## landing

In Martian orbit, the crew will move to a vehicle designed to take them safely to the planet's surface.

### touchdown

32

### exploration

33

# Traveling through space

Astronauts will make the monthslong journey to Mars in a large spacecraft known as a Mars Transfer Vehicle, or MTV. The MTV will have a strong metal exterior to stop air escaping into space and protect everything on board. It will be made of several sections, one of which will provide a place for the crew to live and work. Here they must keep themselves occupied and healthy on the long trip to Mars.

## Weightlessness
The MTV's rockets are mostly switched off as it cruises on a path from Earth to Mars. Everything on board floats, with no force to push objects onto the floor.

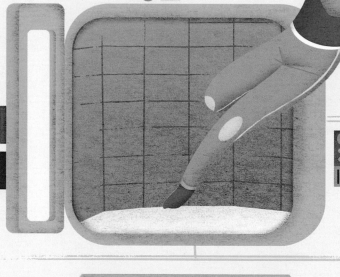

## Sleeping in space
No gravity means astronauts can sleep anywhere—even on the ceiling. Sleep pods give the crew some privacy, but to avoid floating off, they climb into sleeping bags fixed onto the walls.

## Food and drink
The crew's food is freeze-dried into pouches to save space and keep it fresh. Using straws prevents crumbs and droplets from getting into machinery.

## working together

39

## Getting along
Seeing the same people every day for many months could put a strain on the crew. Music, movies, books, and games can provide astronauts with privacy, take their mind off problems, and help team bonding.

## Space medicine
In long periods of no gravity, astronauts' muscles and bones become weaker. They must take a variety of pills and supplements to reduce possible health problems.

## Life on a spaceship
Lights on board follow a 24-hour cycle of day and night. This helps the crew keep an Earthlike routine and adapt to life in space. They are kept busy, but mission planners also make sure they have time to relax.

## Solar threats
With only empty space between them and the Sun, the crew are at risk from harmful rays and blasts of energy called solar flares. Tiny particles can even pass through metal. The MTV is shielded to protect those on board.

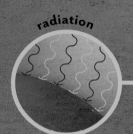

radiation

36

## Regular exercise
Straps holding an astronaut to the ground can help mimic the effects of gravity. They give the crew's muscles a much-needed workout.

### Mars connections
Astronauts must keep fit and healthy during the trip to Mars so they are ready for their mission when they arrive. The spacecraft is fitted with everything they need to survive the long journey.

## Recycle everything!
Water is in limited supply on any long space mission. The MTV's toilet is designed to take water from urine, clean it, and recycle it as drinking water. Solid waste is compressed and stored before being blasted into space.

# Descending to the surface

Once in Martian orbit, the astronauts must get down to the surface of Mars. This is one of the most challenging parts of the mission. The spacecraft they came from Earth on is not built for landing, so the crew must transfer to a special spacecraft called a descent vehicle. This protects them against the heat generated by its rapid journey through the Martian atmosphere. It uses rockets to slow itself so it can touch down gently onto the surface of this new world.

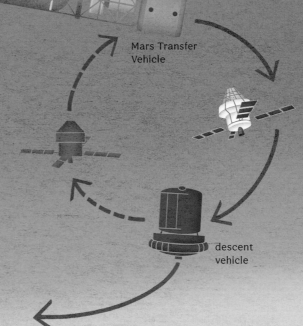

Mars Transfer Vehicle

### Joining the lander
A descent vehicle is sent to Mars orbit before the crewed mission leaves Earth. After the Mars Transfer Vehicle enters Martian orbit by firing its rockets, the crew uses a small space capsule to join the descent vehicle. The capsule then returns to the spacecraft.

descent vehicle

### Martian atmosphere
The air on Mars is thick enough to be a danger to a fast spacecraft, but too thin for parachutes to support the weight of a heavy lander.

### Heat shield
The descent vehicle briefly burns its rocket to slow itself down as it drops toward the surface. A cone-shaped heat shield inflates to protect the lander from getting too hot as it enters the atmosphere.

### Retrorockets
The descent vehicle has slowed down by the time it reaches the lower atmosphere. Large rockets in its base fire downward to slow its descent further.

getting home

### Touchdown
Just above the surface, legs unfold from the base of the descent vehicle. They cushion its landing on the Martian surface.

### On the surface
Pumps deflate the shield and it retracts into the vehicle's base. The lander's legs adjust so that it sits level on the surface.

## Landing on Mars

The descent vehicle is also a living quarters and laboratory, and will be the crew's first base when they land. Other equipment, such as exploration vehicles, is delivered before the astronauts arrive.

### Mars connections

The journey from the spacecraft to the Martian surface needs careful planning. Once they land, the astronauts can begin the first experiments on and explorations of the Red Planet.

### First contact

The astronauts must test their systems and equipment after their descent. They can then step onto the surface of an alien world for the first time in their specially designed spacesuits.

### Angle of approach

To reach the surface safely, the lander must approach at just the right angle. If its entry into the atmosphere is at too steep an angle, it might burn up. If it is too shallow, it may bounce off the atmosphere back into space.

habitat

41

# how would we survive?

Astronauts will have to wear special spacesuits to survive the harsh conditions found on Mars. Because they are too far away to get new supplies from Earth, they will have to take everything they can with them, and find or make anything else they need.

# survival

Earth provides everything we need—fresh air, warmth, and accessible materials. Mars has a hostile environment and its resources are hard to find. Humans must plan carefully if we want to survive there.

## protection

Spacesuits will provide air to breathe and protection from radiation and the cold.

## resources

Astronauts will have to use science and technology to get vital resources.

## transport

A variety of vehicles will be needed for astronauts to explore and find what they need.

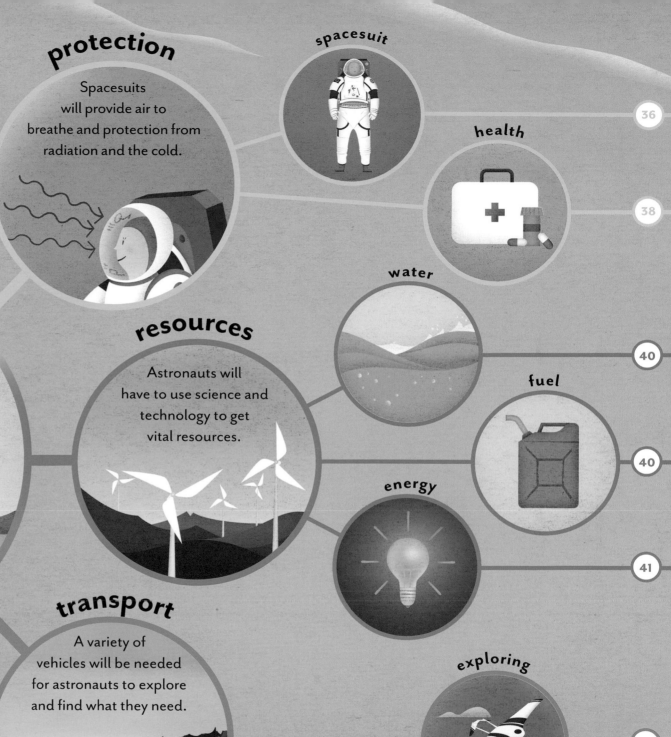

# Protection on Mars

Spacesuits will be vital for humans to survive outside their bases and vehicles on Mars. They keep astronauts warm and safe from dangerous radiation. The design of a suit for Mars will have to be different from any made in the past. It will need to be more lightweight and flexible so people can move easily in Martian gravity. It will also have to be tougher to resist the danger of punctures and other accidents as astronauts work.

## Suit design

A spacesuit for Mars needs to offer protection but also be easy to work in. It should combine a solid helmet and upper body, easy-to-reach life-support controls, and flexible arms and legs. Airtight seals lock everything together.

Strong lights help astronauts work in dark conditions. Cameras record what they see.

The helmet is fitted with microphones and a speaker for easy communication.

### Helmet
A strong plastic dome is the best shape to resist the pressure from air inside the suit. It also protects against any damaging impacts.

A clear dome allows a wide field of vision.

Gloves are strong but flexible for grasping tools and collecting rocks.

## Radiation
Unlike Earth, Mars has a thin atmosphere and no magnetic field. This means it cannot block dangerous ultraviolet rays from the Sun and tiny but fast-moving space particles. Spacesuits are designed to protect astronauts from these types of radiation.

health

The suit filters out poisonous carbon dioxide breathed out by the wearer.

A drinking-water supply leads to a straw inside the helmet.

Pure oxygen gas is pumped into the suit for the astronaut to breathe.

## Toxic atmosphere

The thin Martian air is mostly carbon dioxide, a gas that in large amounts is toxic to humans. It is important that astronauts have a supply of breathable oxygen.

## Life support

Each spacesuit comes with a backpack that provides life support. It supplies the astronaut with clean air, water to drink, and heating or cooling so they can control their body temperature.

terraforming 58

## Temperature extremes

It can reach 68°F on Mars, but temperatures are usually far below freezing. They can even dip to -238°F at night. Warm water pumped through thin tubes helps keep the wearer warm.

### Mars connections

Freezing temperatures, toxic air, and dangerous radiation mean humans cannot survive on Mars without help. Specially designed spacesuits will keep astronauts warm and safe when they are not inside.

# Staying healthy

The long voyage to Mars and conditions on its surface could cause a variety of problems for the astronauts' bodies and minds. With help from Earth so far away, it is vital the crew stays healthy for a successful mission. Anyone chosen for the trip will have been tested for health problems and trained to cope with stress, but doctors at Mission Control on Earth will still need to monitor the crew and give treatments when needed.

### Brain health
Changes to the brain will last for months after astronauts land on Mars and get used to gravity again. This could cause problems such as clumsiness and disorientation.

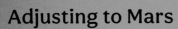

### Adjusting to Mars
Missions to space stations have taught doctors about some of the problems of travel in weightless conditions. This has allowed them to come up with various treatments for when astronauts land on Mars.

### Heart and blood
The muscles that pump blood around the body can get lazy in space. This could make astronauts feel lightheaded when they get to Mars.

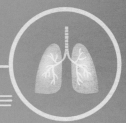

### Lungs
Our breathing systems are one of the few things that adapt well to space travel.

### Bone strength
Sunlight helps make healthy bones. The Sun's rays on Mars are dangerous to humans, so astronauts will need to take vitamin pills to keep bones strong.

### Immune system
Special cells that fight infections get weaker in spaceflight, making it harder to recover from infections or wounds. A well-stocked medicine cabinet will help protect astronauts.

becoming Martian

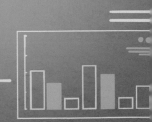

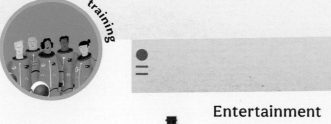

training

## Entertainment
Music, games, digital books, and movies will help astronauts relax. Regular messages from friends and family on Earth will also offer support.

### Mars connections
Being so far from Earth's help means Mars crews cannot afford to get injured or sick if they want to complete a mission smoothly. Longer future missions will need doctors, robots, and computers to support a crew's health.

## Coping with the mission
A long mission to Mars will put a lot of stress on the crew. They will be isolated from Earth but crammed alongside each other, and mistakes could prove disastrous. They will need ways to relax and take the pressure off.

## Work schedules
Mission Control will plan a regular pattern of work, relaxation, and sleep for the crew. As well as science research, there will also be everyday chores such as cooking and cleaning.

## Keeping fit
Exercise on Mars will boost the crew's strength and endurance. It will also reduce their stress levels and help them stay focused.

# Finding resources

Mars has many of the resources humans need to survive there. Ice from the soil can be melted to produce water. As well as being used for drinking, this water can be processed to make oxygen for breathing and fuel for power plants and even rockets. The problem is accessing these resources. To make the most of them, astronauts will have to spend a lot of their time working with some ingenious engineering.

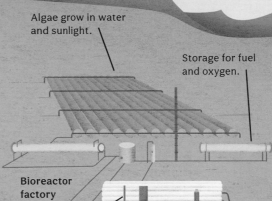

Algae grow in water and sunlight.

Storage for fuel and oxygen.

Bioreactor factory

Concentrators extract chemical waste.

### Growing fuel

Rocket fuel could be made on Mars using tiny plantlike algae from Earth. These grow by taking in carbon dioxide from Martian air. They are then fed to the microbe *E.coli*, which produces the fuel chemical and oxygen as waste.

### Mining water
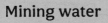

In places, Martian soil is mixed with up to 50 percent ice. This could be accessed by drilling into the soil and heating the surroundings. The melted water would then be pumped to the surface. Water would need careful treatment to make it pure.

### Extracting oxygen
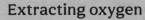

To make oxygen, electricity could be used to split water molecules, or $H_2O$, in sealed tanks. This releases bubbles of oxygen, O, and hydrogen gas, H, which would be collected and stored.

## Electricity

Humans need power from electricity to live and work on Mars. Sunlight and wind can be weak there, so astronauts will have to use a mix of different technologies to make enough power. Equipment for these can be delivered before humans arrive, and even be assembled by robots.

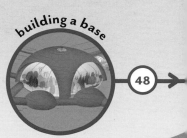

building a base 48 →

Solar panels must be kept clear of Martian dust so they can absorb sunlight.

Turbines need to be lightweight so they can turn in the weak Martian wind to make electricity.

Heat from mini nuclear reactors boils water to make steam. This turns small turbines to generate electricity.

## Habitat

The first astronauts will live in a simple base made of pods brought from Earth. These will provide protection from the hostile Martian conditions. Crews will also have vehicles for exploring and machinery for harvesting resources.

## Construction materials

A large 3D printer can make new buildings as bases get bigger. It uses a mix of heated, crushed Martian rocks and metals brought from Earth.

### Mars connections

Humans can only survive on Mars with access to water, oxygen, energy, and fuel. They cannot take enough of these resources with them from Earth, so must make them from what they find on Mars.

# Transport on Mars

Mars might be smaller than Earth, but it is still a big place to cover. Astronauts will need a mix of transport to find what they need to survive, as well as to explore interesting areas. Wheeled vehicles will allow them to travel long distances and even act as mobile bases for long expeditions. Flying robot drones and remote-controlled rovers will collect information about the surroundings so important areas can be chosen for more detailed investigation.

### Glider drone
Lightweight solar-powered drones carry cameras and other instruments high into the sky to survey the landscape around the astronauts.

### Getting around
For short trips, astronauts can ride on an open buggy. This has large seats for people in spacesuits, as well as storage for equipment and rock samples. Each wheel has its own motors and suspension to roll over rough ground.

### Exploring in comfort
To cover large areas at speed, explorers will use a vehicle with a pressurized cabin. This allows them to wear normal clothes while riding inside.

rovers

18

## Weather balloons
Balloons filled with lightweight gas can rise into the atmosphere and drift for weeks or months. They carry out detailed surveys and even deliver science instruments to hard-to-reach areas.

## Taking to the air
Small helicopter drones with fast-spinning blades make short flights to spot interesting areas for study or potential resources. They also check the terrain ahead for safety.

*exploration*

46

## Living on the move
For long-distance expeditions, astronauts can use large, pressurized vehicles that can carry supplies for several weeks. These include spacesuits, robots, and smaller vehicles.

## Robot assistants
Wheeled robots can help astronauts with a variety of tasks, such as drilling out rock samples, carrying equipment, and filming activities.

## Mars connections
Transport on Mars is vital if astronauts are to learn more about the planet and find the resources needed to survive there. Airborne robots can also travel to give more information about possible exploration sites.

# could we
# live there
# permanently?

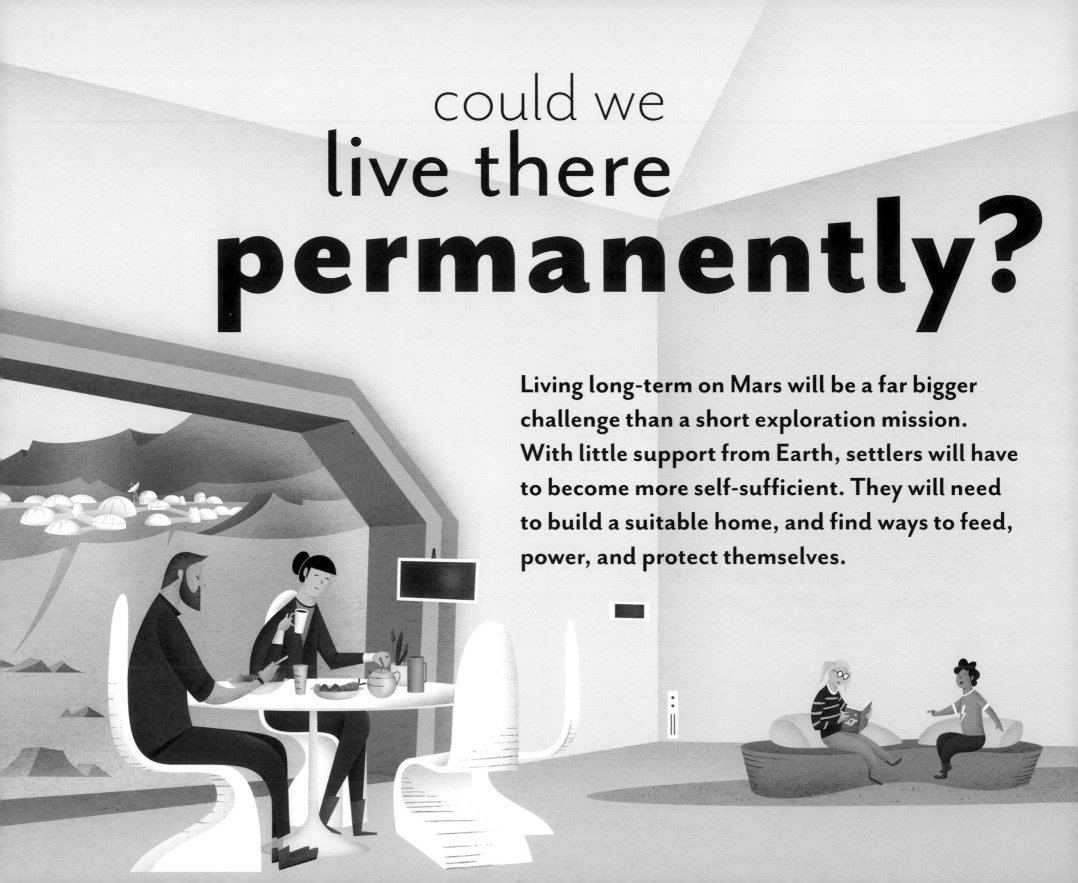

Living long-term on Mars will be a far bigger challenge than a short exploration mission. With little support from Earth, settlers will have to become more self-sufficient. They will need to build a suitable home, and find ways to feed, power, and protect themselves.

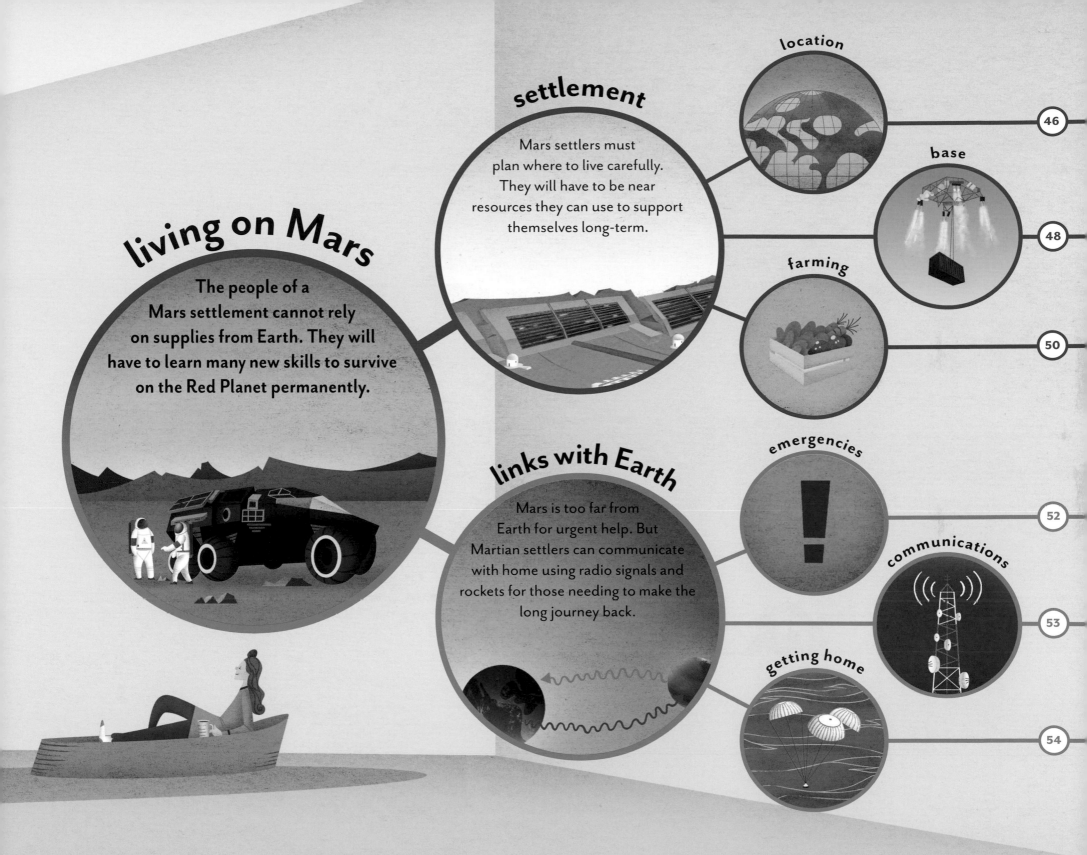

# living on Mars

The people of a Mars settlement cannot rely on supplies from Earth. They will have to learn many new skills to survive on the Red Planet permanently.

## settlement

Mars settlers must plan where to live carefully. They will have to be near resources they can use to support themselves long-term.

location

46

base

48

farming

50

## links with Earth

Mars is too far from Earth for urgent help. But Martian settlers can communicate with home using radio signals and rockets for those needing to make the long journey back.

emergencies

52

communications

53

getting home

54

# Where to settle

Humans will need to find a suitable place to live if they are to stay on Mars permanently. But the question is where. The planet's landscape is varied like Earth's, with volcanoes, craters, canyons, and ice caps. It also has a similar pattern of seasons and changing weather, which could make certain regions challenging to live in. Robot explorers and human missions will need to survey all the options so they can choose the best.

## Wish list
The site for a permanent base must balance lots of needs. It will have to be sheltered, offer access to ice to make water, air, and fuel, and have reliable sunlight for power. It's unlikely that any one location will be perfect.

Mars surface

12

## Searching for sites
Astronauts will have to spend some of their time traveling to sites that could be used as bases. Their observations will help decide the most suitable location for people to live long-term.

## Volcanoes
Most volcanoes on Mars are thought to be extinct. The flow of their ancient molten lava has left behind caves and tunnels. These could be used as shelters for Martian settlers.

## Hot rocks
Rocks around extinct volcanoes might still be hot. Settlers could make energy by pumping cold water through them. This creates steam, which can drive tubines to make electricity.

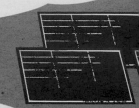

## Highlands

Most of southern Mars has a cratered landscape. There is water ice hidden beneath the red dust, but big seasonal changes mean bitter cold and little sunlight in winter, making it a difficult place for humans to live.

building a base 48

## Craters

The bottom of deep, bowl-shaped craters, such as Hellas Planitia in the southern hemisphere, could be an option for a human base. These types of crater offer extra protection against the radiation on the Martian surface.

## The equator

Areas on the Martian equator are the warmest on Mars, and easier to land on. They also get the most reliable sunlight for generating solar power. Yet they tend to have drier soil with less ice for making water.

## Mars connections

Finding a location for a base with enough sunlight and access to ice for water is vital for humans living on Mars long-term. Some experts think an area between the equator and the north pole might be the safest bet.

# Building a base

A settlement on Mars will need many different types of buildings to support the people living in it. These include housing, greenhouses for growing food, power stations, and factories both for extracting Martian resources and for turning them into useful products. Any building used by people for long periods will have to be protected from Martian conditions, with heating, a supply of breathable air, and shielding from dangerous radiation.

materials

### Launch pad
Rockets for traveling to and from Martian orbit launch and land at a pad outside the crater wall. They are far enough away that any accidents do not harm the settlement.

## Living on Mars
Spaces for living and relaxing might be built into the wall of a crater to shelter people from the Sun's dangerous radiation. Settlers could reach other parts of the base through tunnels covered by Martian soil, which absorbs the solar rays.

## Greenhouses
Plants need as much sunlight as possible for healthy growth. On Mars, food plants would need to be grown in domed greenhouses that can receive light from all sides.

## Mining operations

Water could be pumped from below the surface using wells that melt underground ice. Martian soil could be mined from the surface and processed to get useful chemicals.

### Backup power

A field of windmills would provide backup power. Wind would spin the turbine blades when the Sun is blocked out during Martian dust storms.

## Solar power station

A huge field of mirrors focuses sunlight on a tower to make energy for the base. Heat collected by the tower boils water into steam, which expands to spin turbines and generate electricity.

Solar panels provide a further source of power.

mining water

3D printers could use cement made from Martian soil for new buildings.

60

## Building design

Domes are a good shape for buildings on Mars because they spread pressure evenly over their surface. This would stop buildings from bursting from the high pressure of the air inside compared to the weak surrounding atmosphere.

## Mars connections

A permanent Martian base will need spaces for living, growing food, mining, and making power. All buildings will need to be designed so they protect people from dangerous radiation.

## Avoiding radiation

Particles from the Sun are most dangerous when they come from directly overhead. Many buildings could use a mushroom-shaped design with a thick protective roof for protection. Transparent sides allow light in when the Sun is low down and safer.

# Farming

Settlers on Mars will not be able to bring enough food with them to survive for long. They will need to learn how to farm the Martian surface if they want to support themselves for many years. As well as providing food, plants can also help to break down and reuse various types of waste produced by humans. Over time, this will allow the the settlers to create a sustainable food system similar to that on Earth.

## Tests in orbit
Astronauts have grown plants on the International Space Station. It helped them understand which types of plant can best survive harsh conditions.

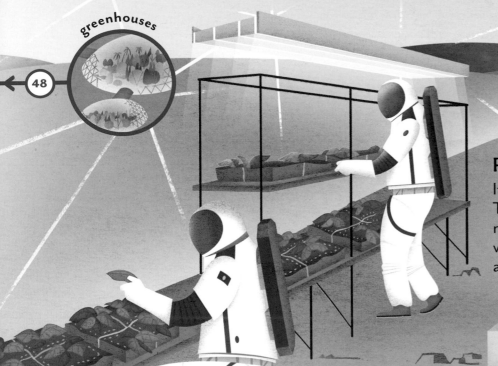

greenhouses

48

## Plant life on Mars
It will not be easy for plants to live on Mars. They grow by absorbing carbon dioxide, the main gas on Mars. But they still need oxygen, which humans will have to make for them, and energy from Mars's weak sunlight.

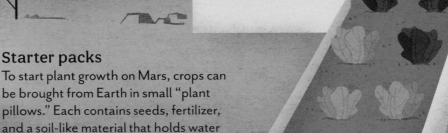

## Starter packs
To start plant growth on Mars, crops can be brought from Earth in small "plant pillows." Each contains seeds, fertilizer, and a soil-like material that holds water supplies and helps plants' roots.

## Crop selection
Settlers will grow fruit and vegetables for the important nutrients humans need. These could include salad leaves, peas, beans, tomatoes, and cereals. Some crops could be altered in laboratories so they can cope better in Martian conditions.

greening Mars

61 →

## Artificial light
Most plants will not grow as large or healthy in Mars's weak sunlight as they would on Earth. Artificial light panels matching the color of Earth's sunlight will help boost their growth.

## Growing in Martian soil
Farming in Martian soil will be difficult. It contains toxic chemicals, and is too thin and dusty to hold water and nutrients. It will have to be specially treated to make it suitable for growing plants.

Farmers wear spacesuits inside to protect them from radiation.

## A vegan diet
There will be no animals on Mars, so humans will not be able to eat meat, dairy, or eggs. Instead, they will have to follow a vegan diet.

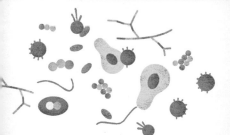

## Cleaning microbes
Microbes called blue-green algae could be brought from Earth and added to the Martian soil. These could remove harmful chemicals and enrich the soil with nutrients.

## Natural fertilizer
Solid human waste can be used as manure to feed plants. Urine could also be used to make a plant-boosting fertilizer.

### Mars connections
Humans will need to grow food on Mars if they are to support themselves. We must make sure there is no existing Martian life that could be put at risk before Earth plants and microbes are allowed to spread.

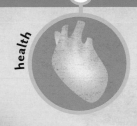

# Emergency!

People living and working on Mars will be isolated from Earth, with any possible rescue mission months or even years away. In emergencies, they may not even have time to wait to get advice from experts on Earth. Fortunately, weak gravity on Mars reduces some risks. Falls are less likely to cause injury, and dropped objects less likely to get damaged. But when accidents do happen, settlers cannot afford to take any chances and will always need to think of safety first.

## Dangers on Mars

Mars does not experience disasters such as major earthquakes or floods, but winds and the dust storms they create can cause damage. Small accidents and breakdowns are also far more dangerous there, as everything has to work properly to keep people safe.

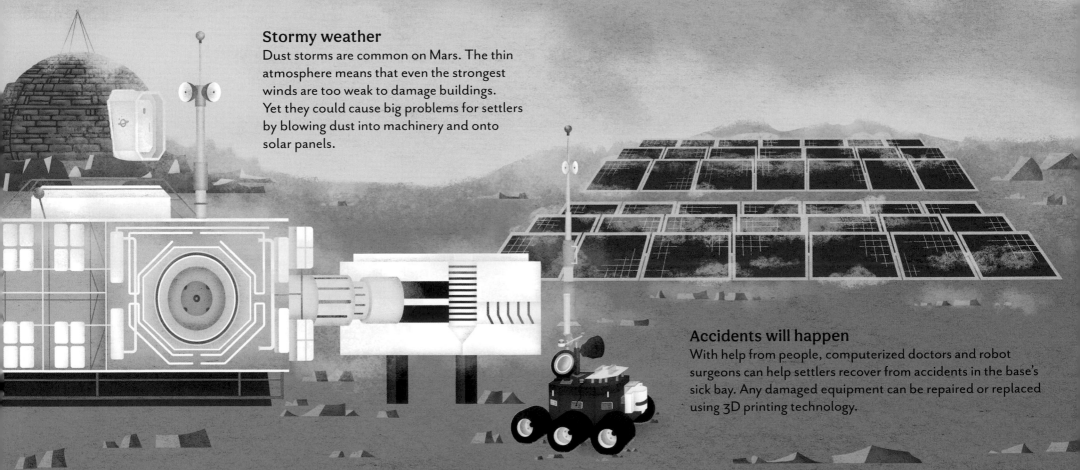

### Stormy weather

Dust storms are common on Mars. The thin atmosphere means that even the strongest winds are too weak to damage buildings. Yet they could cause big problems for settlers by blowing dust into machinery and onto solar panels.

### Accidents will happen

With help from people, computerized doctors and robot surgeons can help settlers recover from accidents in the base's sick bay. Any damaged equipment can be repaired or replaced using 3D printing technology.

## Out of reach
About every two years, Earth and Mars lie on opposite sides of the Sun. This means communication with Earth is impossible, so settlers spend a couple of weeks totally cut off from their home planet.

Activity from the Sun can disrupt radio links to Earth.

Radio signals can be sent directly to Earth, or bounced off satellites in Mars orbit.

## On their own
Contact with Earth from Mars is not straightforward like telephone or messaging. In an emergency, settlers will have to rely on their own wits to solve any problems.

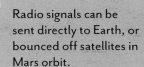

*getting home*

## Communicating with Earth
The settlers stay in touch with Earth using radio signals carrying sound, video, and internet data. Quick chats are difficult, because radio signals take between three and 22 minutes to travel each way.

**Mars connections**

Mars settlers cannot plan for every possible accident, but they will have the tools and equipment to fix what they can. Engineers can try to help from Earth, but in an emergency the settlers will be on their own.

# Getting home

Even if people can start to live permanently on Mars, some trips back to Earth will still be necessary. Just like the journey to Mars, the return to Earth is complex, with several stages. Astronauts first board a special Mars Ascent Vehicle to get off the surface and back into Martian orbit. Then they change to a much bigger spacecraft for the journey home. The final stage is a fiery descent through Earth's atmosphere in a small reentry capsule.

30

space travel

Mars Ascent Vehicle

$H_2O$

$CO_2$

$O_2$

### Refueling on Mars
Sending a Mars Ascent Vehicle from Earth with enough fuel to launch it back to Mars orbit is expensive and risky. Instead, people can process Martian ice to make new fuel.

### Mars Ascent Vehicle
A rocket will take astronauts to Mars orbit to meet the vehicle that will carry them to Earth. It will be automatically landed and be ready for when it is needed. Mars's weak gravity means it does not need to be as big as rockets that launch astronauts from Earth.

Mars
Transfer
Vehicle

## Into orbit
Astronauts will move to a Mars Transfer Vehicle after an ascent vehicle has taken them into orbit. The transfer vehicle fires its engines to leave Martian orbit and begin the monthslong trip to Earth.

Reentry
capsule

**Mars connections**

Even when a Martian settlement is more established, trips back to Earth might sometimes be needed. This helps the planets keep close communications so they can share information, ideas and resources.

## Earth reentry
A small reentry capsule leaves the Mars Transfer Vehicle and plummets toward Earth with the crew on board. Collisions with gas in the upper atmosphere heat the capsule's base to thousands of degrees, so it is specially built to withstand the heat.

## Arriving safely
Falling through Earth's thick atmosphere slows the capsule, before parachutes open to slow it further. A splashdown in the sea provides a soft landing for the astronauts on board.

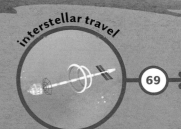

interstellar travel

69

## Astronaut return
The crew emerge from their capsule and await rescue by a support ship. After years on Mars, their muscles will struggle with Earth's stronger gravity. A team of medical experts will need to monitor them to help them recover.

# would humans change **Mars?**

People in the future will find it easier to live on Mars if conditions are more Earthlike. Transforming the whole planet could take centuries, over which time humans will also start to change. Yet perhaps one day people will be able to walk on the surface without spacesuits.

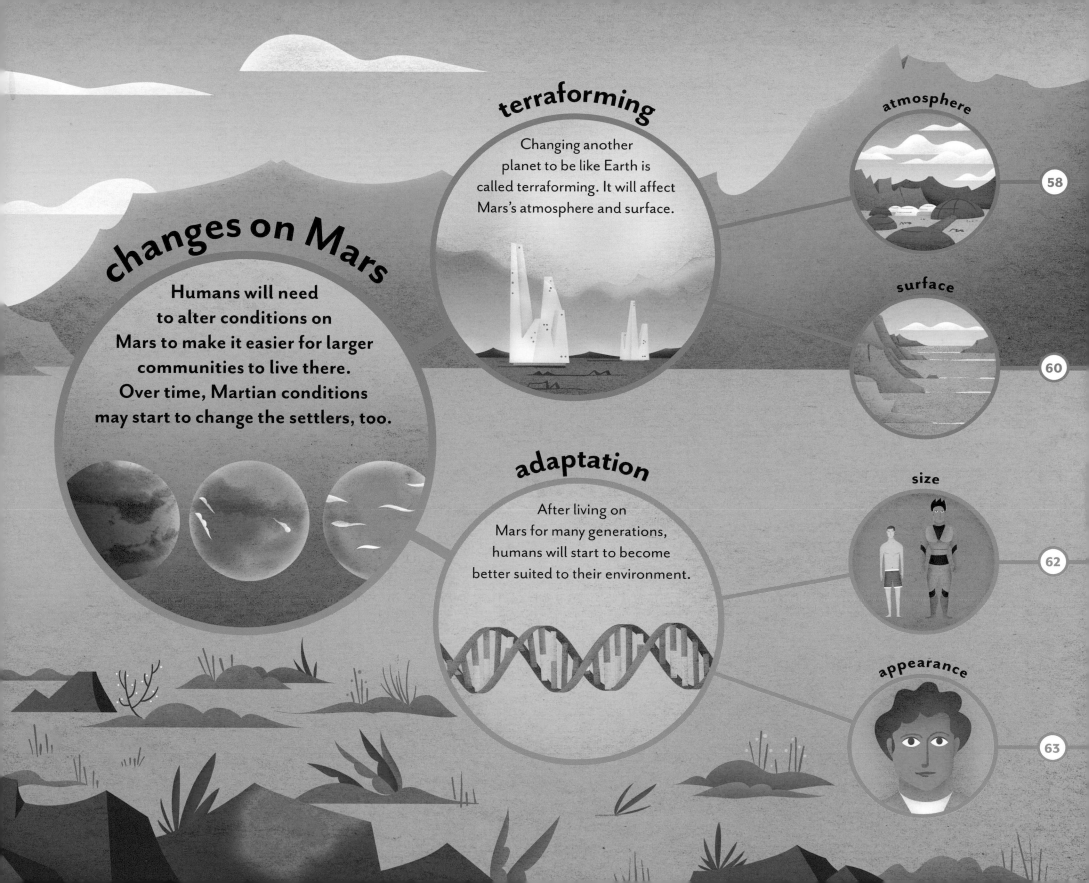

# changes on Mars

Humans will need to alter conditions on Mars to make it easier for larger communities to live there. Over time, Martian conditions may start to change the settlers, too.

## terraforming

Changing another planet to be like Earth is called terraforming. It will affect Mars's atmosphere and surface.

## adaptation

After living on Mars for many generations, humans will start to become better suited to their environment.

atmosphere

58

surface

60

size

62

appearance

63

# Thickening the atmosphere

Mars's atmosphere is mostly toxic carbon dioxide gas–
$CO_2$–and is far too thin to trap heat around the planet to
warm the cold, rocky surface. Settlers must instead rely on
heavy protective spacesuits. Yet thickening the atmosphere
by adding much more gas could help to warm Mars up. This
would make it much easier for people to live and work, even
if they still could not breathe the air. There are a few ways
humans could try to change Mars's atmosphere.

## Delivery from space
Astronauts could use rockets to nudge
icy asteroids or comets so they hit Mars.
This could release their large amounts
of ammonia gas to help thicken the
atmosphere. But it could be dangerous
for Martian settlers.

## Making an atmosphere

Extra gas is needed to build a Martian
atmosphere. It could be made from raw
materials found on Mars, released more
slowly by changing the Martian surface,
or be delivered from outside sources.

## Mobile factories
Huge rolling factories could
mine rocks from the surface.
These would be used in
chemical reactions that pump
out gases called CFCs. These
are very good at trapping heat
in a planet's atmosphere.

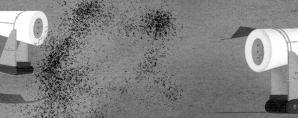

## Darkening the planet
Mars's polar caps contain huge amounts of $CO_2$ ice.
Spraying sooty chemicals onto them would cause
them to absorb more heat from the Sun. This raises
their temperature and releases $CO_2$ gas, thickening
Mars's atmosphere.

greening Mars

61

## A second Sun
Massive mirrors in space
could be used to double
the Sun's heating power.
By reflecting sunlight onto
Mars's poles to melt the ice,
more $CO_2$ is released into
the air.

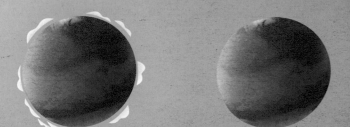

solar threats

## Losing an atmosphere

Billions of years ago, Mars had a thick atmosphere, which acted like a blanket to give it a warm surface. As the power of the Sun blew the atmosphere into space, the Martian surface froze.

### Solar wind

Tiny particles escape from the Sun at very high speeds and blow out across the Solar System. This is the solar wind.

Deimos

Phobos

### Magnetic shield

A giant satellite orbiting between Mars and the Sun could help the planet keep its atmosphere from being blown away by the solar wind. The satellite would create an artificial magnetic field that would deflect solar wind particles safely around Mars.

artifical magnetic field

A magnetic shield casts a "shadow" across space in which Mars is protected from the solar wind.

## Keeping an atmosphere

To keep hold of a thicker atmosphere, Mars would need to be shielded from a stream of particles from the Sun that could strip away its atmosphere. Mars lacks a strong magnetic field that does this job for Earth.

**Mars connections**

If future inhabitants of Mars are going to live on a warm planet, they will first need to thicken its atmosphere. This will rely on technology we have yet to develop.

# A second Earth

The first step in terraforming Mars to be more like Earth would be to thicken its atmosphere to make the planet warmer. This would melt ice and cause rivers, lakes, and seas of liquid water to form. The next stage would be to carefully treat the soil to remove toxic chemicals and add nutrients. This could allow simple plants and other microscopic life-forms from Earth to survive on Mars. These would slowly make the air breathable for humans and animals.

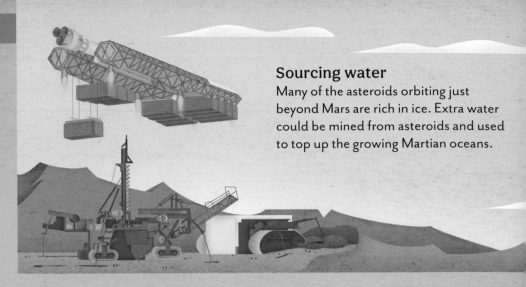

### Sourcing water
Many of the asteroids orbiting just beyond Mars are rich in ice. Extra water could be mined from asteroids and used to top up the growing Martian oceans.

### Is there enough water?
Even if all of Mars's ice was melted, it would only produce about a tenth of the water that existed on the planet in the past. This might not be enough to start a water cycle like the one that supports life on Earth.

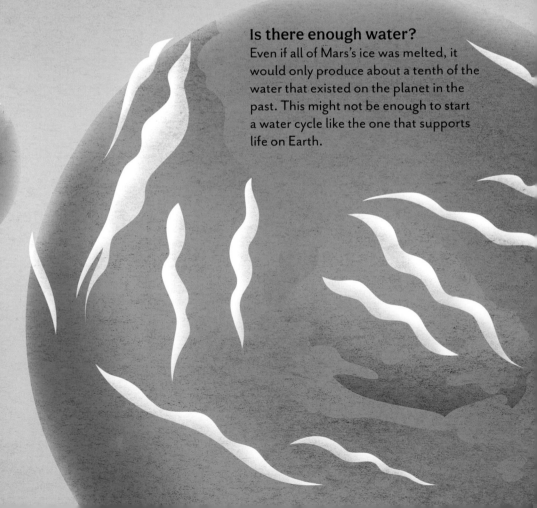

### Blue planet
If humans can make Mars warmer, water will begin to flow from the melting polar caps and emerge from the deep-frozen soil. It will collect in low-lying areas, such as crater bottoms and the northern plains, creating lakes and oceans.

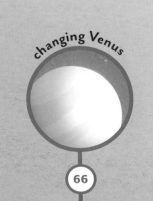

changing Venus

66

11

life on Mars

## Planetary protection

Before terraforming, scientists will need to confirm Mars has no life of its own. Today's space probes are specially treated to make sure no microbes from Earth contaminate Mars, which is sure to happen in terraforming.

## Martian meteorites

Scientists study rocks that have fallen to Earth from Mars to look for signs of life. If they find any, it will mean we need to rethink what we do on Mars, as it could harm any life still there.

## Martian water cycle

After many thousands of years, Mars could become a warm, wet planet. Its terraformed atmosphere would absorb water vapor rising from its seas, then drop it back onto the land as rainwater to support plant life.

## Greening the surface

The first step in turning Mars green would be to introduce microscopic blue-green algae. These simple life-forms would absorb carbon dioxide from the atmosphere and replace it with oxygen. More complex plants that need oxygen to survive could then be introduced.

### Mars connections

Mars must have liquid water on its surface and support plant life to be truly like Earth. Changing the atmosphere to achieve this would take many human lifetimes.

# Becoming Martian

Over thousands of years, people whose families have lived on Mars for generations might start to look different from people on Earth. This is because, over time, some settlers will be born with slight differences that happen to suit them better to their Martian environment. These differences are then spread down family lines. One day, Martian humans could have changed to be perfectly adapted to life on Mars, but would struggle if they ever went to Earth.

## A new species?

Humans today all belong to a single species, or group, called *Homo sapiens*. If Martian settlers change so much that they could no longer have children with Earth-born humans, they would be classed as a new human species.

health

38

## Body shape

Martians might develop a sturdier skeleton with chunkier limbs. These would help make up for having weak bones.

## Immunity

Martians would not be exposed to all the germs found on Earth. Their immune systems might be weaker, leaving them at greater risk from diseases carried by visitors from Earth.

## Bones

Mars's weak gravity will make the bones of people born on Mars lighter and more spongelike compared to those of Earth humans. This means they would be more easily broken.

## Bigger heads
Some scientists think that over time people on Mars might develop bigger heads. But this would not necessarily make them smarter!

## Life undercover
Before Mars is terraformed, settlers will spend most of their lives sheltering indoors. This could at first make them shorter and paler due to weaker muscles from lower Martian gravity and less sunlight.

## Adapted vision
Settlers' eyes might become more sensitive to cope with the lower levels of sunlight on Mars. Their pupils or entire eyes might get bigger to let more light in.

## Orange skin
Darker skin could help protect people from the damaging effects of the Sun's radiation. Some experts think that settlers' skin might turn orange after Mars is terraformed.

### Mars connections
Over many generations of living on Mars, people might evolve to be more suited to Martian conditions. Some scientists think space radiation could speed up this process.

# where
# can we
# go next?

Once humans are established on Mars, the Solar System has many other worlds to explore. And if one day we can overcome the vast distances between stars, there are billions of other planets spread across our galaxy that humans could call home.

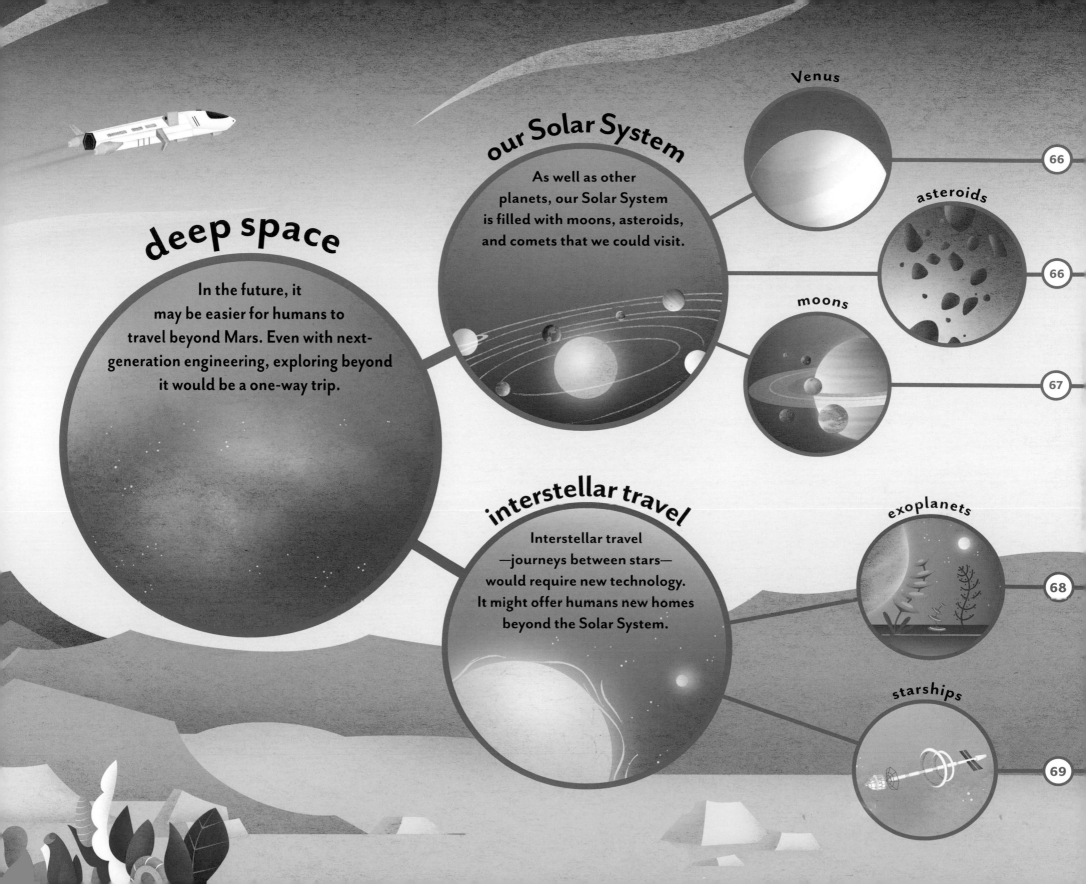

# deep space

In the future, it may be easier for humans to travel beyond Mars. Even with next-generation engineering, exploring beyond it would be a one-way trip.

## our Solar System

As well as other planets, our Solar System is filled with moons, asteroids, and comets that we could visit.

Venus

66

asteroids

66

moons

67

## interstellar travel

Interstellar travel —journeys between stars— would require new technology. It might offer humans new homes beyond the Solar System.

exoplanets

68

starships

69

# Across the Solar System

Using twenty-first-century engineering, it would take humans many years to visit other worlds in the Solar System. But future technologies might make it easier to reach these places. Scientists think the best destinations to explore are further out, beyond Mars. Here, in the extreme cold and fading light of a distant Sun, people could visit asteroids and moons with underground oceans. These could be suitable for new human bases and be a chance to search for alien life.

terraforming

58

## Exploration targets

Our Solar System contains hundreds of planets, moons, asteroids, and comets, and there are plenty of reasons to visit. Some might tell explorers more about our own origins, or even be home to other life-forms.

### Ceres

Beyond Mars in the asteroid belt is the dwarf planet Ceres. It has volcanoes that ooze salty ice. Scientists think these salt deposits could contain simple life-forms.

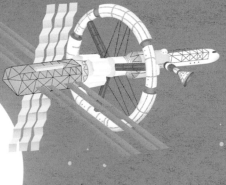

### Venus

Some scientists think Venus could one day be terraformed to become an Earthlike home for us. Today, its crushing atmosphere and ovenlike temperatures make human visits impossible.

### Asteroids

Some asteroids contain vast amounts of useful metal we could mine. Others have not changed since the birth of the Solar System, so could reveal vital information about its beginnings.

Saturn has more than a hundred moons, as well as a beautiful ring system.

## Titan

Saturn's biggest moon is the only other place in the Solar System that has a thick atmosphere and liquid lakes on its surface. Its Earthlike nature could tell us about conditions on Earth before life evolved.

Titan has rain and lakes of oily chemicals rather than water.

Ganymede

## Enceladus

This small moon of Saturn has oceans of liquid water just below a thin ice layer. Water escapes into space in jets that burst through cracks in the ice, before freezing and falling back to the ground as snow.

Jupiter has no surface to land on, and radiation belts that could be deadly for explorers who get too close.

## Europa

This large moon of Jupiter has a thick, icy crust covering an ocean of water, kept liquid by undersea volcanoes. Explorers could send robots through the crust to look for life.

Europa may be the most likely place to find life elsewhere in the Solar System.

### Mars connections

Setting up bases on worlds with similarities to Earth may help us find alien life. Some scientists think ancient comet impacts delivered starter kits of chemicals needed for life to many worlds in our early Solar System.

# Into the stars

Traveling beyond the Solar System involves crossing immense distances. The closest stars are tens of thousands of times farther away than Mars, and could take centuries to reach even with advanced future technology. These journeys will be one-way trips to start new colonies, or settlements, on other worlds. The final destination for these interstellar voyages would be planets with Earthlike conditions orbiting other stars within our galaxy.

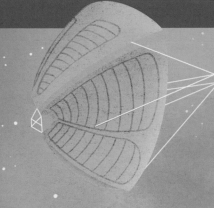

### Sailing by laser
One option for interstellar travel is to shine a powerful laser beam from Earth onto a huge, mirrorlike sail. This would slowly push a spaceship to very high speeds. It would only need engines to slow down at its destination.

## Exoplanets
Planets orbiting other stars are called exoplanets. They are even more varied than the planets in our Solar System, and at least some probably have conditions suitable for supporting human settlers.

### Interstellar distances
The distances between stars are so huge that it will take many lifetimes to reach a destination. Humans making the trip from Earth might never see their new home.

An umbrella-like shield protects the bulk of the ship from collisions with interstellar dust.

too hot

just right

too cold

Venus

Earth

Mars

## The Goldilocks zone
Like Earth, an exoplanet suitable for humans would orbit a star in a Goldilocks zone. This is a region where temperatures are not too hot, not too cold, but just right for liquid water to survive on the surface. Mars is on the edge of our Solar System's Goldilocks zone.

Even hostile exoplanets could have habitable moons for humans.

Settlers live inside giant rings. These slowly spin to produce the effects of gravity.

The ship is driven by controlled nuclear explosions rather than a chemical rocket.

## Colony ships

Traveling to another star would take several lifetimes. An interstellar spaceship would have to be a huge self-contained world, with everything needed to support generations of human colonists.

## Living on a spaceship

The inside of a colony ship would be a mini version of Earth. It would have an atmosphere, water, plants, animals, and even an artificial Sun designed to keep conditions liveable and produce all the supplies needed by the human crew.

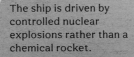

Mars connections

Other worlds in our Solar System cannot support humans long-term without terraforming. Spending lifetimes traveling to Earthlike exoplanets offers a chance for humans to prosper elsewhere.

# Glossary

**algae** Small plantlike living things that make energy using carbon dioxide and sunlight.

**asteroid** A small, rocky object on its own orbit around a star.

**astronomy** The study of stars, planets, and other objects in space.

**atmosphere** The layer of gases and clouds that surrounds a planet.

**booster rocket** A rocket on the side of a launch vehicle that produces power for the first few minutes of flight.

**carbon dioxide** A gas that humans create as a waste product. The atmosphere of Mars is 95 percent carbon dioxide, which is toxic to humans.

**colony** A permanent settlement on another world, populated by people called colonists.

**descent vehicle** A vehicle designed for putting humans safely on the surface of another world.

**exoplanet** A planet outside of our Solar System, usually orbiting another star.

**Goldilocks zone** The area around a star with the right temperature for liquid water to exist on any orbiting planets.

**gravity** A force of attraction between different objects. It makes things fall to the ground on Earth.

**habitat** A small building on another world. Astronauts can live there without a spacesuit.

**heat shield** The part of a spacecraft that resists heat when it enters a planet's atmosphere at high speed.

**inner Solar System** The area of the Solar System between the Sun and the asteroid belt.

**interstellar travel** Travel beyond the Solar System to other stars.

**lander** A spacecraft designed to touch down on the surface of another world.

**launch vehicle** A vehicle designed to carry a smaller vehicle away from the surface of a planet and into space.

**Lunar Gateway** A planned space station in orbit around the Moon. It could be used as a departure point for trips to Mars.

**magnetic field** The area around a magnetic object in which some other objects are pushed or pulled by a force called magnetism. Some planets have magnetic fields.

**Mars** The fourth planet from the Sun. It is the easiest for humans to explore.

**Mars Ascent Vehicle** A vehicle designed to carry people from the surface of Mars to an orbiting Mars Transfer Vehicle.

**Mars Transfer Vehicle** A spacecraft designed to carry astronauts on the journey from an Earth or Moon orbit to Mars orbit and back.

**microbe** A tiny living thing that can only be seen under a microscope.

**Mission Control** Where Earth-based scientists and engineers monitor space missions.

**moon** A natural satellite of a planet. Mars has two moons—Phobos and Deimos.

**nuclear reactor** A machine that generates energy by using the forces inside atoms.

**opposition** When two planets are at their closest point to each other on the same side of a star.

**orbit** The path that an object in space takes around a planet or star.

**orbiter** A crewed or robot spacecraft that stays in orbit to study a planet.

**outer Solar System** The area of the Solar System beyond the asteroid belt.

**oxygen** A gas needed by almost all animals on Earth to survive.

**radiation** In space, energy that travels through space as light rays or tiny particles. Higher-energy forms of radiation can be harmful to most living things.

**reentry capsule** A small spacecraft designed to return astronauts from space safely to Earth's surface.

**resources** Things useful to humans, such as raw materials, air, water, and energy.

**robot explorer** Machines designed to explore other worlds. Some stay in orbit, while others land on a planet's surface.

**rocket** An engine that burns chemicals to power a vehicle's motion, or a vehicle that uses this type of engine.

**rover** A robotic wheeled or tracked vehicle for exploring another world.

**satellite** Any object that orbits another one. A satellite can be natural, such as a moon, or human-made.

**settler** Someone who sets up a new life elsewhere and helps to make the settlement more liveable.

**Solar System** The region of space containing the Sun and everything that orbits it.

**solar flare** Bursts of energy on the Sun that send dangerous radiation across the Solar System.

**solar panel** A panel covered in small devices that make electricity from sunlight.

**solar power** Energy from the Sun that can be collected for use by humans, using devices such as solar panels or mirrors.

**solar wind** A stream of particles blown from the surface of the Sun across the Solar System.

**space probe** A spacecraft designed to explore space and other planets, and controlled from Earth.

**spacesuit** Clothing that protects an astronaut from conditions in space or on the surface of other worlds.

**telescope** A device for making distant objects appear nearer.

**terraforming** Transforming another world so that it is more like Earth.

**3D printing** A process that uses machines to build up materials in layers to create three-dimensional objects.

**turbine** A propellerlike device that produces electricity when it spins.

**water ice** Frozen water. It could be melted for use by astronauts on Mars.

**wind turbine** A windmill-like tower that uses large blades to make electricity as it spins in the wind.

# Index

Author: Giles Sparrow
Illustrator: El Primo Ramón
Consultant: Dr. Elizabeth Rampe, NASA

Editor: George Maudsley
Designer: Claire Cater

Published by EarthAware Kids
Created by Weldon Owen Children's Books
A subsidiary of Insight International, L.P.
PO Box 3088
San Rafael, CA 94912
www.insighteditions.com

Insight Editions:
Publisher: Raoul Goff
Senior Production Manager: Greg Steffen

ISBN: 979-8-88674-162-9

Printed in China
First printing July 2024  DRM0724

10 9 8 7 6 5 4 3 2 1

FSC
www.fsc.org
MIX
Paper | Supporting
responsible forestry
FSC® C188448